Energy
133

冰冻术和缩骨功

Freeze or Squeeze

Gunter Pauli

[比] 冈特·鲍利 著

[哥伦] 凯瑟琳娜·巴赫 绘

贾龙智子 译

上海远东出版社

丛书编委会

主　任: 田成川

副主任: 闫世东　林　玉

委　员: 李原原　祝真旭　曾红鹰　靳增江　史国鹏

　　　　梁雅丽　孟小红　郑循如　陈　卫　任泽林

　　　　薛　梅　朱智翔　柳志清　冯　缨　齐晓江

　　　　朱习文　毕春萍　彭　勇

特别感谢以下热心人士对童书工作的支持:

匡志强　宋小华　解　东　厉　云　李　婧　庞英元

李　阳　梁婧婧　刘　丹　冯家宝　熊彩虹　罗淑怡

旷　婉　王靖雯　廖清州　王怡然　王　征　邵　杰

陈强林　陈　果　罗　佳　闫　艳　谢　露　张修博

陈梦竹　刘　灿　李　丹　郭　雯　戴　虹

目录

Contents

一只蟑螂正在西伯利亚游历。它想去拜访火蜥蜴——这位朋友冬天会陷入冬眠，只有来年春天阳光普照的时候才会苏醒。

"谢谢你大老远地来看我，我知道这可是一段很漫长的旅途。"火蜥蜴说。

A cockroach is visiting Siberia. It wants to meet the salamander that freezes in the winter – only to wake up in the spring again when the sun shines.

"Thank you so much for coming all this way. I know it was a very long trip just to see me," says Salamander.

一只蟑螂正在西伯利亚游历。

A cocroach is visiting Siberia.

这里的冬天实在太冷了……

It gets very cold in winter …

"别介意，我随时都可以过来。我不是很喜欢我住的地方，所以能受到这么热烈的欢迎真是太开心啦。这可从来没发生过。"蟑螂回答道。

"我听说了。但是来这儿的人没有能待得长的。这里的冬天实在太冷了，最低能到零下50℃！除非你能在自己的血液里产生糖分，不然会冻死的。"

"Well, I can come anytime. I am not very well liked where I live, so it is a joy to be welcomed so warmly. This has never happened before," responds Cockroach.

"I heard. But anyone who comes here can't stay for long. It gets very cold in winter, as cold as minus 50 degrees Celsius! You will freeze to death unless you make sugars in your blood."

"糖？这么说为了在这种能冻死人的天气里活下来，必须把自己变甜？"蟑螂问道。

　　"哦是的，我的火蜥蜴和蛙类朋友们都能制造糖。我们把细胞里所有水分都排出去，然后在天气变得特别冷的时候陷入酣睡之中。"

"Sugar? So you must turn sweet to survive in this icy cold weather?" Cockroach asks.

"Oh yes, my salamander and frog friends all make sugar. We push all the water out of our cells and then fall into a deep sleep when it becomes too cold."

......陷入酣睡之中......

... fall into a deep sleep ...

你的心脏也会停止跳动吗？

Does your heart stop?

"你的心脏也会停止跳动吗？"

"一切都停滞了，身体的一切都不运行了。我相当于变成了一块冰。"

"Does your heart stop?"

"Everything stops, nothing works. I turn into a block of ice."

"那如果我把你扔到地上，会发出小石子敲击地板的声音吗？"

　　"是的，我看上去就像一块石头一样，这种状态会持续好几个月，甚至好几年。"

"So if I were to drop you on the ground would it sound like a rock hitting the floor?"

"Yes, I do look like a rock for months, and sometimes even for years."

......看上去就像一块石头一样......

... look like a rock ...

……蟑螂的生命力也很顽强。

... cockroaches have a great life force as well.

"你的生命力太顽强啦！我知道很多人会想向你学习。"

"我听说你们蟑螂的生命力也很顽强。"

"You must have a great life force. I know many who would love to learn from you."

"I hear that cockroaches have a great life force as well."

"我们必须如此，因为差不多所有人都想抓住我们，杀死我们。"

　　"但你们族群还是顽强地存在了一亿多年！看起来没人成功地消灭了你们。"火蜥蜴说。

"We need it. Just about everyone is after us to kill us."
"Yet you have been around for more than a hundred million years. Seems like no one has ever been successful in eradicating you," Salamander says.

......所有人都想抓住我们，杀死我们。

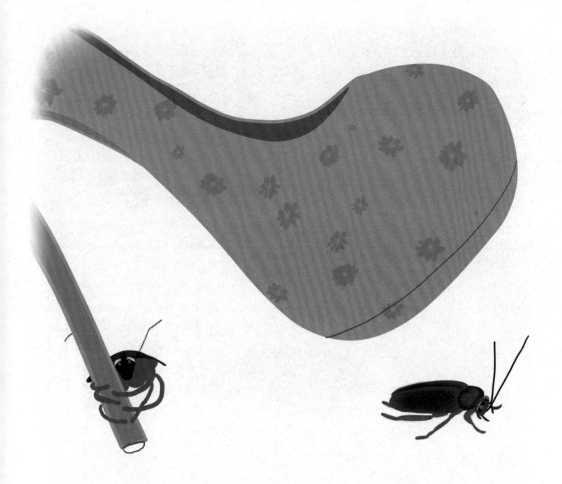

... everyone is after us to kill us.

我们速度很快。

We are very fast.

"这是真的。我们速度很快，短短一秒之内跑动的距离就可以是身体长度的50倍！而且我们可以挤过2毫米宽的小洞。"

"那么说说看，你是怎么能在一小时里跑200英里，并且把身高从5毫米缩到2毫米的？"

"True. We are very fast. We run fifty times our body length in just one second! And we can squeeze through a hole that is only two millimetres wide."

"But tell me, how do you manage to run two hundred miles an hour, and reduce your height from five to two millimetres?"

"嘿，你会冰冻术，我会缩骨功啊！我只用把身体展平然后把腿伸出去。"

"你可真聪明！你知道吗，如果我们按别人的期望生活，我们可能都活不下去了。我们都是真正的幸存者！"

……这仅仅是开始！……

"Hey, you can freeze, I can squeeze. I just flatten my body, and stretch my legs out."

"How clever of you! You know, if we behaved the way people expect us to, neither of us would still be around. We are true survivors!"

… AND IT HAS ONLY JUST BEGUN! …

······ 这仅仅是开始！ ······

··· AND IT HAS ONLY JUST BEGUN! ···

After being frozen for years many metres down in the permafrost, the Siberian salamander, also known as a newt, can thaw out and then start moving and run away. Frozen salamanders that have fallen into cracks have been found in layers of ice 14 metres down.

西伯利亚火蜥蜴又名蝾螈，可以在数米之下的冻土层里冰冻数年后解冻并自己回到陆地上来。人们曾在14米以下的冰层中发现跌落裂缝的被冰冻的火蜥蜴。

The salamander lives in Siberia, close to the Kolyma River where it can survive in temperatures as low as minus 50° C. It has no amphibian competitors north of the Arctic Circle, except the Siberian frog.

火蜥蜴的栖息地在西伯利亚，它们能够在靠近科雷马河温度低至零下50℃的地方存活。除了西伯利亚蛙，它们在北极圈以北没有两栖动物竞争对手。

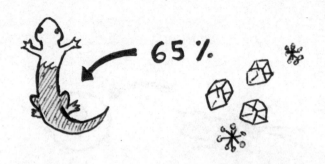

在西伯利亚，夏季只有三到四个月，除此以外的其他月份里火蜥蜴都保持冬眠状态。它们身体内部高达65%的水分会变成冰。比起被冻死，火蜥蜴更可能缺水致死。

The summer lasts only three to four months, and for the rest of the year the salamander remains frozen. Up to 65% of the water in its body becomes ice. Salamanders are more often killed by lack of water than by the freezing cold.

$$CH_2-CH-CH-CH-CH-CHO$$
$$\quad OH \quad OH \quad OH \quad OH \quad OH$$

葡萄糖

火蜥蜴能够制造由葡萄糖和甘油组成的"防冻剂"，用于取代血液和细胞中的水分并保护组织不被锋利的冰晶所伤。

The salamander produces "antifreeze" consisting of glucose and glycerol that replace water in blood and cells and protect tissue against damage caused by sharp ice crystals.

Cockroaches fit through tiny spaces by flattening their skeleton and splaying their legs to the side.

蟑螂通过展平躯干，把腿展开到身体两侧来通过狭小的空间。

Roaches can run fast, while flattening themselves. They run at a rate of 1.5 metres (or 50 body lengths) per second. For a human, that would be like running 200 miles per hour.

蟑螂身体放平的时候跑动速度很快，可达每秒1.5米（相当于其身长的50倍）。这个速度对人类来说，相当于每小时跑200英里。

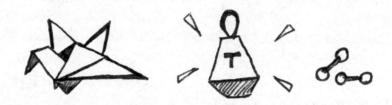

蟑螂的身体可以像折纸一样折叠起来。然而，即使在其坚硬的外骨骼上施加很大的压力（比其体重大100倍），蟑螂仍然可以跑得跟以前一样快，看不出任何受伤的痕迹。

A roach's body can fold like origami. However, even when great pressure (100 times its body weight) is placed on its rigid exoskeleton, the cockroach can still run as fast as before without any signs of being harmed.

蟑螂筑巢并为幼虫寻找食物。它们以鸟粪为食，从中获得足够的氮。蟑螂会清除幼虫巢中的死蟑螂和真菌。这种亲代抚育的行为对昆虫来说是很不寻常的。

*C*ockroaches build nests and find food for their larvae. They eat bird droppings to get enough nitrogen. Roaches sweep their nurseries of dead cockroaches and fungi. Such parental care is unusual among insects.

If a frog and a salamander can freeze and then thaw and wake up again after months of being frozen like a block of ice, could we do the same one day?

如果青蛙和火蜥蜴能够把自己冻起来，并在长达数月的冰冻状态后解冻并苏醒过来，我们将来也能做到吗？

If freezing and thawing is possible due to a sugar, what more could be achieved with sugar apart from eating it?

如果冰冻和解冻是由于糖分的缘故，那么除了食用以外，糖还能用来做什么？

What would you do to squeeze through a tiny space?

为了挤过一个狭小的空间，你会怎么做？

When you think of a cockroach, do you imagine it to be a clean and caring insect? How is it that we have such a negative impression of this insect?

当你想到蟑螂时，你会想象这是种干净而有爱心的昆虫吗？我们是怎么对这种昆虫抱有如此负面的印象的呢？

Do It Yourself!

自己动手！

Can you squeeze through tiny spaces? Let's try. Do not do this on your own though; make sure that there are at least two people to help you, in case you get stuck. Find a small opening, at home or at school, and try squeezing through it. Imagine you have lost your key and the only way to get in is through a small window, or a pet entrance. Search the web for some techniques that will allow you and your friends to get through spaces that at first sight seem impossibly small. Now discuss your experience of squeezing through like a cockroach with your friends.

你能从狭小的空间通过吗？让我们来试试吧。不要单独进行此项活动，确保至少有两个人协助你，以免你被卡住。在家里或学校里找一个小洞，尝试挤过去。想象你丢了钥匙，进门的唯一方式就是通过小窗或是宠物入口。在网上搜索一些技巧能够让你和朋友们通过那些第一眼看来小得不可思议的空间。现在跟朋友们讨论一下像蟑螂一样挤过狭小空间的体验吧。

学科知识
Academic Knowledge

生物学	蟑螂跟白蚁属于同一种族，有共同的祖先；现存的蟑螂多达4 000种，属于蜚蠊科双翅目；蟑螂有社会结构，彼此之间会沟通并能够识别自己的后代；每只蟑螂有三对足，每只足上有五个爪子；蟑螂通过腹腔进行呼吸；蟑螂表现出群体行为；蟑螂进行集体决策；火蜥蜴的栖息地是高山森林，它属于冷血动物，通过冬眠度过严冬。
化 学	蟑螂用甘油避免结晶，从而能够在冰点下存活；西伯利亚火蜥蜴使用葡萄糖和甘油来避免细胞脱水和萎缩；防冻剂的作用；盐可用于防冻，但具有腐蚀性；甲醇、乙二醇、丙二醇和甘油是无腐蚀性的防冻剂。
物 理	蟑螂的抗辐射能力是人类的6至15倍；敏捷性和灵活性的结合；运动定律；仅仅通过折叠，将一张普通的方形纸变成容器、动物或花草。
工程学	蟑螂是CRAM（具有铰接装置的可压缩机器人）机器人的原型，该种机器人在受力时仍可移动，可承受高达其体重20倍的压力；折纸技巧被用于开发车载安全气囊和支架植入。
经济学	居住在长期处于冰点以下地理区域的群体需要将水管隔热，并施用防冻剂以避免水管损坏，因此增加了成本；折纸的原理被应用在支架、打包和工程应用方面，以减少成本和原料使用，同时提高了部署速度。
伦理学	在只有其中少数个体被视为问题的情况下妖魔化整个族群（比如蟑螂）。
历 史	罗马历史学者老普林尼记录了蜚蠊在多种药物中的应用；蟑螂是第一个在太空繁衍后代的地球物种（2007年在一艘俄罗斯宇宙飞船上）。
地 理	蟑螂能够在北极的寒冷和热带的炎热下存活；西伯利亚占俄罗斯大陆的77%，从乌拉尔山脉向东延伸到太平洋和北极，到哈萨克斯坦的丘陵和蒙古国及中国的边界。
数 学	蟑螂每小时跑动距离可达5千米，考虑到其身体长度，相当于人类每小时跑330千米；人们通过折纸来证明几何构型。
生活方式	蟑螂在人类留下污垢的地方繁衍生息，它们长大的时候会蜕皮，脱下的废弃物是引发哮喘的过敏源；我们是在解决问题的结果还是在解决其根源所在？
社会学	蟑螂（cockroach）一词来自西班牙语cucaracha；在西方，蟑螂是令人厌恶的，然而在墨西哥和泰国则是一种食物来源，在中国则用作药物，且蟑螂养殖场的数量不断增加；折纸（orugami）一词源于日语，其中oru指的是"折叠"，而gami指的是"纸"。
心理学	过度泛化：为什么所有的蟑螂都被认为是害虫，即使其中绝大多数都无害且在生态系统中发挥重要作用（4 000种蟑螂中仅有4种可归类为害虫）？
系统论	大自然有非凡的能力，能够调整、自我进化并创造出在最极端条件下仍然能够存活的成分。

情感智慧
Emotional Intelligence

火蜥蜴

火蜥蜴在蟑螂抵达的时候表示了感激，感谢它的拜访。火蜥蜴对蟑螂的遭遇表示同情，但也指出它所生活地区的严寒条件不允许长期逗留。火蜥蜴分享了它的解决办法，即使这是种革命性的突破方法，火蜥蜴仍然很谦逊，自嘲说把自己变成了冰块。火蜥蜴好奇心强，希望能多了解蟑螂的生命力。火蜥蜴告诉蟑螂自己非常欣赏它，明白蟑螂在如此多外界的杀意下克服了重重困难存活下来。火蜥蜴发现了蟑螂在速度和力量方面展现出的非凡能力，询问了更多细节。火蜥蜴并没有立即对蟑螂做出回应，而是注意细节和自己的观察。它用哲学的方式总结道，动物解决问题的方式会让人类大吃一惊。

蟑　螂

蟑螂知道几乎没有人类喜欢它，因此对于火蜥蜴带给它的热情欢迎和体验感到很高兴，并且充满了感激。它不相信用糖分维持细胞以应对严寒天气的办法，也不相信这种天然的化学物质会让火蜥蜴的身体停止一切活动，包括心跳。蟑螂愿意与热衷于了解如何在不利条件下保持健康和活力的人分享它的智慧。蟑螂承认很多人追着想要消灭它们，但没人成功过。它有自知之明并且很自信，愿意解释它的行为。比起讲述过多的细节，它开玩笑说自己会"缩骨功"。

艺术
The Arts

日本折纸是一种折叠纸张的艺术，不用剪刀或胶水即可将平整的纸张变成三维形体，是一种非常自然的造物方式。折纸是一种需要遵循简单原理的艺术。了解一下折纸艺术，学习基本的原理，成为一名折纸艺术家吧。给你的父母和朋友，甚至将来某天给你的孩子看看你的纸艺作品。

思维拓展
Systems: Making the Connections

大自然为生态系统提供食物、物质与能量。生物多样性提供了适应不断变化的环境的力量。这种灵活性大于适应性，这种力量即使在人类无法想象能够存活下来的恶劣环境下，仍然能创造出维持生命所需的成分。其导致的生物学和化学现象很容易被经验不足的人当作是奇迹。

防冻剂对于很多行业来说是一种必需品。防冻剂中大部分化学物质是石油化工产品，具有毒性。而防冻剂的高性能天然替代品不仅容易获取，还可以在环境中轻而易举地处理掉。

西伯利亚火蜥蜴在艰苦的生存条件下，利用糖分作为防冻剂，它们非同一般的表现被人们广泛肯定，与之不同的是，蟑螂有能力达到类似的成果，却几乎从未被人们提及。人类似乎忘记了蟑螂之所以在各种杀虫措施下仍然存活和繁衍生息，是因为人类在蟑螂种群重新出现的地方留下了太多的垃圾。人类将蟑螂贬为地球上最卑劣的生命，忽视了它们的社会行为、它们对后代的奉献、它们的亲属关系以及保持清洁的程度。所以我们不仅应该重新思考我们生产的用以制造冷冻剂的化学物质，还应该想想我们是怎样对待那些我们不知道或不了解的生命的。

动手能力
Capacity to Implement

列一个可获取的防冻剂产品清单。如果你所生活地区的气候冬天非常寒冷，你可能已经对此有所了解了；如果你住在热带地区，可能从来没想过这些。不过，防冻剂也用于引擎的冷却剂，因此即使不用作防冻剂，仍然是可以获得的。研究那些日常使用的防冻剂，查证哪些具有毒性，哪些能够接触皮肤，以及哪些种类甚至人体能够摄入而不会对健康造成危害。基于这些分析，列出你可能更愿意使用的产品的特征：技术表现、对环境的影响，以及最重要的，对人类生命的风险性。

故事灵感来自

This Fable Is Inspired by

莎南·斯迈利
Shanan Smiley

莎南·斯迈利女士是一位自然保护生物学家，拥有美国蒙大拿州立大学的生物学学位。她获得了佛蒙特州绿山学院的理学硕士学位，这所大学被认为是世界上最好的可持续发展大学之一。莎南作为生物学家的职业生涯起始于美国林务局，对巢穴、濒危植物和害虫进行调查。莎南在纽约州莫霍克保护区时任自然保护生物学家。她承担综合害虫管理研究并对蟑螂进行了研究。莎南向小学生解释气候变化和物种现象，撰写科普性的文章，并在美国国内和国际的科学及教育工作者会议上发表她的作品。她也定期出席当地的电视、广播和国家的公共广播节目。莎南还是户外摄影爱好者和野外消防员。

图书在版编目（CIP）数据

冈特生态童书.第四辑：修订版：全36册：汉英对照 /
（比）冈特·鲍利著；（哥伦）凯瑟琳娜·巴赫绘；
何家振等译.—上海：上海远东出版社，2023
书名原文：Gunter's Fables
ISBN 978-7-5476-1931-5

Ⅰ.①冈… Ⅱ.①冈…②凯…③何… Ⅲ.①生态环
境−环境保护−儿童读物—汉、英 Ⅳ.①X171.1-49

中国国家版本馆CIP数据核字（2023）第120983号
著作权合同登记号图字09-2023-0612号

策　　划　张　蓉
责任编辑　曹　茜
封面设计　魏　来　李　廉

冈特生态童书
冰冻术和缩骨功
[比]冈特·鲍利　著
[哥伦]凯瑟琳娜·巴赫　绘
贾龙智子　译

记得要和身边的小朋友分享环保知识哦！
八喜冰淇淋祝你成为环保小使者！